# Healthful Snacks

Everybody likes to eat snacks, but some snacks are more healthful than others.

How can we make healthful and tasty snacks? Let's read this card. We can **add** the food together to make a healthful snack.

**Fruit Salad**

6 grapes
3 pieces of pineapple

Mix together.

Grapes and pineapple make a healthful snack.

We **add** the fruit together to make a fruit salad.

$$\begin{array}{r} 6 \\ + 3 \\ \hline \end{array}$$

This is a healthful snack that is made with cereal and raisins.

We **add** the pieces together to make trail mix.

### Trail Mix

7 pieces of cereal
4 raisins

Mix together.

4

$$7$$
$$+\ 4$$

Fruit juice is a **liquid** that can be made into a fruit pop.

A fruit pop is a healthful snack.
This is how you can turn it into a **solid** snack!

## Fruit Juice Pops

Pour juice into the cups and put them in the freezer.

Add sticks when the juice is almost frozen.

# Can you count the number of juice pops?

How many snacks did we make **altogether**?

Which healthful snacks do you want to try?

There are a lot of healthful snacks.